QUANTUM JOURNEYS
Billy Carson's Exploration of Teleportation

Oteren.Fredrick

DISCLAIMER

The data gave in this book is for instructive and educational purposes as it were. While each work has been made to guarantee the exactness and unwavering quality of the substance, the writer makes no portrayals or guarantees of any sort, express or inferred, about the fulfillment, precision, dependability, appropriateness, or accessibility as for the book or the data, items, benefits, or related illustrations contained in this.

COPYRIGHT

© 2024 [Oteren.Fredrick]. All Privileges Reserved.

No piece of this book might be duplicated, put away in a recovery framework, or sent in any structure or using any and all means, electronic, mechanical, copying, recording, etc., without the earlier composed consent of the writer.

TABLE OF CONTENT

INTRODUCTION

CHAPTER 1:
- The Foundation of Teleportation
- Hypothetical Underpinnings of Instant transportation
- Contextual investigations and Hypothetical Models

CHAPTER 2:
- Billy Carson's Quantum Insights
- How Quantum Hypotheses Backing the Chance of Instant transportation
- Key Investigations and Discoveries via Carson

CHAPTER 3:
- The Mechanics of Teleportation

CHAPTER 4:
- Real-world Applications
- Contextual analyses and Experimental runs Projects

CHAPTER 5:
- Case Studies and Experiments
- The 1997 Innsbruck Investigation

- The 2004 NIST Trial
- The 2012 Chinese Quantum Satellite Analysis
- Billy Carson's Outstanding Contextual investigations
- Cognizance Driven Trials

CHAPTER 6:
- Future of Teleportation
- Quantum Incomparability and Instant transportation
- Combination with Quantum Organizations

CHAPTER 7:
- Philosophical and Spiritual Dimensions
- The Idea of The real world and Presence

CHAPTER 8:
- The Intersection of Science Fiction and Reality
- Sci-fi's Effect on Instant transportation Exploration
- Genuine Motivations from Fiction
- Overcoming any barrier: From Fiction to Plausibility
- Expected Leap forwards and Advancements

CHAPTER 9:
- Challenges and Controversies
- Quantum Intelligibility and Decoherence

CONCLUSION
APPENDICES
ACKNOWLEDGEMENTS

INTRODUCTION

Outline of Billy Carson and His Work
Billy Carson is an eminent figure in the fields of old civic establishments, power, and cutting edge innovations. Known for his profound experiences into the secrets of the universe, Carson has committed quite a bit of his vocation to investigating ideas that challenge traditional logical comprehension. Among his various advantages, transportation stands apart as an especially interesting theme. Carson's work overcomes any issues between old insight and present day science, offering an interesting viewpoint on the capability of human capacities.

Prologue to the Idea of Instant transportation
Instant transportation, the speculative exchange of issue or energy starting with one point then onto the next without crossing the actual space between them, has long enamored the creative mind of researchers and aficionados the same. Promoted by sci-fi, the idea proposes the chance

of prompt travel, changing comprehension we might interpret reality. While frequently consigned to the domain of imagination, ongoing progressions in quantum mechanics and hypothetical material science allude to the achievability of such peculiarities.

Verifiable Points of view on Instant transportation

The possibility of instant transportation is certainly not a cutting edge creation. Old texts and folklores from different societies reference types of momentary travel or bilocation, where people show up in various places all the while. Accounts of divine beings, spiritualists, and shamans performing marvelous accomplishments have been gone down through ages, proposing a well established interest with the idea. In the advanced time, transportation has been investigated broadly in sci-fi writing and media, laying the preparation for serious logical request.

Billy Carson's Mission for Instant transportation

Billy Carson's advantage in instant transportation is profoundly entwined with his more extensive journey to grasp the secret capability of human cognizance and the universe. Drawing from a different scope of sources, including old messages, quantum material science, and his own visionary encounters, Carson presents a convincing case for the believability of instant transportation. His work dives into the mechanics of quantum snare, the job of awareness in forming reality, and the likely uses of instant transportation in regular daily existence.

The Meaning of Instant transportation Exploration

The investigation of instant transportation isn't only a scholarly activity; it holds significant ramifications for the fate of humankind. Effective advancement of instant transportation innovation could reform transportation, correspondence, and, surprisingly, the manner in which we see reality itself. It challenges how we might interpret the basic laws of physical science and opens up new roads for logical and mechanical progression.

Construction of the Book

"Quantum Excursions: Billy Carson's Investigation of Instant transportation" is organized to give an extensive outline of Carson's experiences and examination. Starting with essential ideas in quantum mechanics, the book advances through the hypothetical and useful parts of instant transportation, genuine applications, and the philosophical ramifications. En route, perusers will experience contextual investigations, test results, and theoretical future situations, all outlined inside Carson's interesting viewpoint.

Toward the finish of this excursion, perusers will acquire a more profound enthusiasm for the capability of instant transportation and the visionary work of Billy Carson. This investigation isn't just about the study of being anyplace; it's tied in with growing the skylines of human comprehension and opening additional opportunities for our aggregate future.

CHAPTER 1:

The Foundation of Teleportation

Quantum Mechanics Rudiments
Quantum mechanics, the part of material science that arrangements with the way of behaving of particles at the littlest scales, frames the bedrock of current instant transportation speculations. At its center, quantum mechanics depicts how particles like electrons and photons display both wave-like and molecule like properties, a peculiarity known as wave-molecule duality. This duality prompts the idea of superposition,

where a molecule can exist in various states at the same time until noticed.

Another key standard is quantum entrapment, where coordinates or gatherings of particles become connected so that the condition of one molecule right away impacts the condition of the other, paying little mind to separate. This "creepy activity a ways off," as Albert Einstein broadly named it, is a vital component in the hypothetical underpinning of instant transportation.

Hypothetical Underpinnings of Instant transportation

Instant transportation, as conjectured in the domain of quantum mechanics, includes moving the condition of a molecule starting with one area then onto the next without truly moving the actual molecule. This cycle depends on snare and a system known as quantum state move.

The essential thought is to make a caught sets of particles, one situated at the takeoff point (Alice) and the other at the objective (Sway). A third molecule (Charlie), whose state is to be magically transported, communicates with Alice's molecule. Through a progression of quantum estimations and traditional interchanges, Bounce's molecule assumes the territory of Charlie, successfully finishing the instant transportation.

While this interaction at present applies to quantum states as opposed to plainly visible articles, it makes way for understanding how more mind boggling types of instant transportation could be created.

Effects on Billy Carson's Instant transportation Hypotheses

Billy Carson's advantage in instant transportation is impacted by a rich embroidery of logical, philosophical, and otherworldly sources. His interest with old civic establishments and their implied progressed information assumes a critical part in forming his speculations. Texts from old Egypt, Sumeria, and different societies frequently portray instant transportation like peculiarities, recommending that our predecessors might have perceived parts of reality that advanced science is simply starting to investigate.

Also, Carson draws from contemporary logical exploration in quantum mechanics and awareness studies. He sets that the actual psyche might assume a vital part in empowering instant transportation, recommending that human cognizance and quantum mechanics are personally connected.

Key Ideas in Instant transportation Exploration
1. Quantum Entanglement: The interconnection between particles that permits quick data move.
2. Superposition: The capacity of particles to exist in different states all the while, falling into a solitary state upon estimation.
3. Quantum State Transfer: The method involved with moving the condition of a

molecule starting with one area then onto the next utilizing trap and traditional correspondence.

Contextual investigations and Hypothetical Models

A few momentous investigations have shown parts of quantum instant transportation. One of the most prominent is the 1997 investigation by specialists at the College of Innsbruck, who effectively magically transported the quantum condition of a photon across a distance of a few kilometers. This investigation affirmed the possibility of quantum state move, giving a proof of idea to future examination.

Billy Carson's hypothetical models frequently integrate these trial discoveries while developing them with bits of knowledge from antiquated intelligence and supernatural encounters. He imagines a future where instant transportation isn't restricted to particles yet can be applied to bigger items and, surprisingly, people, on a very basic level changing comprehension we might interpret space, time, and reality.

Understanding the underpinning of instant transportation requires an excursion through the multifaceted and frequently strange universe of quantum mechanics. By joining logical standards with experiences from old civic establishments and supernatural investigations, Billy Carson offers an interesting viewpoint on this entrancing peculiarity. In the accompanying sections, we will dive further into Carson's quantum experiences, the mechanics of instant transportation, and the expected applications and ramifications of this progressive innovation.

CHAPTER 2:

Billy Carson's Quantum Insights

Carson's Understanding of Quantum Trap

Billy Carson sees quantum snare as something other than a curious element of quantum mechanics; he considers it to be an extension between the physical and powerful domains. Ensnarement, as indicated by Carson, is proof of a more profound, interconnected reality that rises above conventional actual limits. He accepts that this interconnectedness can be outfit to accomplish instant transportation, at the quantum level as well as possibly for bigger articles and, surprisingly, people.

Carson proposes that old developments could have had a comprehension of snare and involved it in manners that cutting edge science is simply starting to uncover. His understanding goes past established researchers' ordinary view, recommending that cognizance itself could assume a fundamental part in keeping up with these caught states.

How Quantum Hypotheses Backing the Chance of Instant transportation

Quantum speculations, especially those including entrapment and superposition, lay the basis for the hypothetical chance of instant transportation. In customary instant transportation conventions, an item's quantum state is estimated, obliterated in the first area, and reproduced in the new area. This interaction depends on the standards of entrapment and traditional correspondence to move the state data without moving the actual matter.

Carson contends that this interaction could be stretched out past the tiny scope, recommending that in the event that we can control caught states and keep lucidness over longer separations and bigger scopes, transportation of plainly visible items could become achievable. He draws on trial proof from quantum instant transportation of particles and photons to help his cases, featuring how these essential analyses approve his more extensive speculations.

Key Investigations and Discoveries via Carson

Billy Carson's work in instant transportation is educated by both hypothetical examination and functional trial and error. He has led different psychological tests and cooperative examinations to investigate the limits of instant transportation. A few vital region of his exploration include:

1. Extended Entanglement: Exploring how entrapped states can be kept and reached out over bigger separations and more intricate frameworks.
2. Consciousness and Quantum States: Analyzing the job of human awareness in affecting quantum states, suggesting that psychological concentration and expectation

could balance out and coordinate ensnared particles.

3. Ancient Techniques: Breaking down authentic texts and antiquities for proof of antiquated instant transportation works on, recommending that lost advances or information could offer bits of knowledge into present day instant transportation research.

Coordinating Old Insight with Present day Science

One of Carson's novel commitments is his mix of old insight with current logical standards. He sets that numerous old developments had progressed information on quantum standards, which they encoded in their fantasies, images, and advancements. For instance, Carson deciphers the Egyptian idea of "ka" (soul or life force) as an old comprehension of quantum states and energy.

By joining these antiquated experiences with state of the art quantum research, Carson intends to open additional opportunities for instant transportation. He accepts that cutting edge science can profit from rethinking antiquated texts and practices from the perspective of quantum mechanics, possibly uncovering lost strategies that could propel current instant transportation innovation.

Difficulties and Contentions
Carson's hypotheses are not without debate. Pundits contend that his understandings of quantum mechanics and antiquated intelligence are speculative and need thorough logical approval. The standard academic local area frequently sees his work with suspicion, especially in regards to the job of cognizance in quantum processes and the functional achievability of naturally visible instant transportation.

In spite of these difficulties, Carson keeps on pushing the limits of regular reasoning, supporting for a liberal way to deal with investigating the secrets of quantum mechanics and instant transportation.

Billy Carson's quantum experiences offer a provocative and extensive perspective on instant transportation, mixing logical hypothesis with otherworldly and verifiable viewpoints. By investigating the interconnected idea of reality through quantum trap and old insight, Carson gives an interesting structure to figuring out the capability of instant transportation. As we dig further into his hypotheses and analyses, we will reveal the mechanics of how instant transportation could function and the significant ramifications it holds for what's to come.

CHAPTER 3:

The Mechanics of Teleportation

The Science Behind Instant transportation: From Hypothesis to Practice

Instant transportation, with regards to quantum mechanics, includes moving the quantum condition of a molecule starting with one area then onto the next without genuinely moving the actual molecule. The cycle relies on quantum snare, where two particles become entwined so that the condition of one straightforwardly impacts the condition of the other, paying little heed to remove. This section digs into the specialized parts of how instant transportation hypothetically and for all intents and purposes works.

Quantum State Move

At the core of quantum instant transportation is the exchange of a molecule's quantum state. The technique starts with making a snared sets of particles, say particles A and B. Molecule A is shipped off the area of the molecule to be magically transported (we should call it C), while molecule B is kept at the objective.

1. Entanglement Creation: Particles A and B are snared.
2. Interaction and Measurement: Molecule A interfaces with molecule C, and a joint estimation of their joined state is performed. This estimation successfully obliterates the first condition of molecule C while delivering traditional data.
3. Classical Communication: The estimation result is shipped off the objective by means of traditional correspondence channels.
4. Reconstruction: Utilizing the got data, the condition of molecule C is remade onto molecule B, finishing the instant transportation.

Innovative Necessities and Progressions
Accomplishing pragmatic instant transportation requires beating huge specialized difficulties. Key innovative headways vital for instant transportation include:

1. Quantum Entrapment Generation: Creating dependable techniques for making and keeping ensnared particles over lengthy separations.

2. Quantum Estimation and Control: Exact estimation procedures to guarantee the precise exchange of quantum states.

3. Error Correction: Carrying out quantum mistake adjustment to moderate the decoherence and loss of data during the interaction.

4. Classical Correspondence Integration: Quick and secure correspondence channels for communicating the estimation results.

Late advancement in quantum figuring and quantum correspondence has carried us closer to acknowledging reasonable instant transportation. Explores different avenues regarding photons and particles have exhibited effective quantum instant transportation over progressively bigger distances, laying the foundation for future turns of events.

Difficulties and Constraints

While critical steps have been made, a few difficulties and limits stay chasing viable instant transportation:

1. Decoherence: Keeping quantum cognizance over lengthy separations and broadened periods is a significant obstacle, as collaborations with the climate can disturb the ensnared state.
2. Scalability: Expanding quantum instant transportation from individual particles to bigger, more perplexing frameworks, including perceptible articles and living creatures, presents significant hardships.
3. Energy and Resources: The cycle requires significant energy and refined assets, making enormous scope execution testing.
4. Classical Correspondence Speed: The need for old style correspondence presents a speed limit, as data can't travel quicker than light.

Likely Arrangements and Advancements
Scientists are investigating different answers for address these difficulties:

1. Advanced Quantum Networks: Creating hearty quantum networks that can keep trapped states over lengthy separations with negligible decoherence.

2. Quantum Repeaters: Utilizing quantum repeaters to expand the scope of entrapment by intermittently invigorating the snared state.
3. Hybrid Systems: Consolidating different quantum frameworks, like photons and iotas, to use their particular assets.
4. Metamaterials and Shielding: Using progressed materials to safeguard quantum states from natural obstruction.

Exploratory Shows and Confirmations of Idea
Various exploratory shows have approved the hypothetical groundworks of quantum instant transportation:

1. Photon Teleportation: Specialists have effectively magically transported the quantum condition of photons over distances as much as a few kilometers utilizing optical strands and free-space correspondence.
2. Ion Trap Experiments: Transportation explores different avenues regarding caught particles have shown high-loyalty quantum state

move, adding to the improvement of quantum registering and correspondence innovations.

3. Quantum Spots and Strong State Systems: Progress in strong state frameworks, for example, quantum specks, has shown guarantee for adaptable quantum organizations and instant transportation gadgets.

The mechanics of instant transportation, grounded in the standards of quantum mechanics, uncover a captivating and complex cycle that mixes hypothetical class with viable difficulties. By understanding the science behind quantum state move and the innovative progressions required, we can see the value in the steps made towards accomplishing instant transportation. As we keep on pushing the limits of what is conceivable, Billy Carson's bits of knowledge and visionary methodology will direct us in investigating the maximum capacity of this progressive innovation. The following parts will dig into this present reality applications, future prospects, and philosophical ramifications of instant transportation, expanding on the basic mechanics talked about here.

CHAPTER 4:

Real-world Applications

Expected Utilizations of Instant transportation in Different Enterprises
Instant transportation, whenever acknowledged on a viable scale, can possibly upset various ventures. From transportation and correspondence to medical care and coordinated factors, the ramifications are huge and groundbreaking.

Transportation
Instant transportation could on a very basic level modify the manner in which we move individuals and merchandise. Immediate travel would wipe out the requirement for customary

vehicles, lessening gridlock, contamination, and the time expected for movement. This could prompt:

- Intercontinental Travel: In a flash move from one side of the world to the next, making global travel as basic as venturing through an entrance.
- Space Exploration: Transportation could work with human space travel, empowering speedy and safe vehicle to and from far off planets or space stations.

Correspondence

Instant transportation of data, especially quantum data, can upgrade correspondence advances:

- Quantum Internet: Secure, quick correspondence networks in light of quantum entrapment, giving extraordinary degrees of safety and speed.
- Information Transfer: Prompt and secure exchange of a lot of information across huge

distances, upsetting ventures dependent on quick, secure data trade.

Medical services
The clinical field stands to benefit gigantically from instant transportation advancements:

- Telemedicine: Quickly transport clinical experts or hardware to remote or underserved regions, further developing admittance to quality medical care.
- Organ Transplants: Transporting organs or natural materials could save incalculable lives by guaranteeing opportune conveyance and decreasing the gamble of corruption.

Planned operations and Inventory network The executives
Instant transportation could smooth out and advance stock chains:

- Immediate Delivery: Merchandise could be magically transported straightforwardly from distribution centers to buyers, reforming web based business and diminishing the requirement for actual transportation organizations.

- Stock Management: Ongoing instant transportation of stock can kill delays and upgrade stock levels across different areas.

Military Applications

The military could bridle instant transportation for different vital and strategic benefits:

- Quick Deployment: Transport troops and gear promptly to basic areas, improving reaction times and functional adaptability.
- Secure Communications: Use quantum instant transportation for secure, carefully designed correspondence channels.
- Observation and Surveillance: Transportation could empower progressed surveillance missions, considering speedy inclusion and extraction of reconnaissance gear or faculty.

Moral and Cultural Contemplations

While the expected advantages of instant transportation are massive, there are critical moral and cultural contemplations to address:

- Protection and Security: Guaranteeing that instant transportation advances are not abused for reconnaissance or unapproved admittance to private spaces.
- Financial Disruption: The appearance of instant transportation could disturb existing ventures and occupation markets, requiring insightful administration of monetary changes.
- Value and Access: Guaranteeing that instant transportation innovation is open and helpful to all, forestalling a split between the people who can manage the cost of it and the individuals who can't.

Contextual analyses and Experimental runs Projects

Investigating expected certifiable applications through experimental runs projects and contextual analyses can give significant experiences:

1. Medical Instant transportation Pilot: A program testing the instant transportation of clinical supplies to distant facilities in underserved districts.
2. Quantum Correspondence Network: Laying out a model quantum web for secure, prompt correspondence between research organizations.
3. Logistics Optimization: Joining forces with significant internet business organizations to direct instant transportation based conveyance frameworks, dissecting proficiency and consumer loyalty.

Cultural Effect

The far reaching reception of instant transportation innovation could prompt significant cultural changes:

- Metropolitan Planning: Urban communities may be overhauled to oblige instant transportation center points, decreasing the requirement for broad transportation foundation.
- Social Exchange: More straightforward and moment travel could cultivate more prominent social trade and understanding, advancing worldwide solidarity.
- Work and Lifestyle: Transportation could empower more adaptable work game plans, lessening the requirement for actual driving and permitting individuals to live further from metropolitan focuses.

The likely certifiable uses of instant transportation are immense and sweeping, promising to change various parts of our lives. From improving medical care and correspondence to upsetting transportation and planned operations, the ramifications of instant transportation are significant. Nonetheless, understanding these applications requires cautious thought of moral, cultural, and mechanical difficulties. As we investigate these applications further, Billy Carson's visionary bits of knowledge and interdisciplinary methodology will keep on directing us towards a future where instant transportation turns into an indispensable piece of our daily existences. The following sections will dig into future progressions, philosophical aspects, and the more extensive effect of instant transportation on how we might interpret reality.

CHAPTER 5:

Case Studies and Experiments

Definite Gander at Fruitful Instant transportation Investigations

In the domain of quantum instant transportation, various examinations have been directed to show and approve the standards hidden this peculiarity. Here, we analyze probably the most critical and effective trials that have made ready for future progressions.

The 1997 Innsbruck Investigation

One of the milestone tests in quantum instant transportation was directed in 1997 by a group drove by Anton Zeilinger at the College of Innsbruck. They effectively magically transported the quantum condition of a photon across a distance of a few meters. This trial included three principal steps:

1. Entanglement Creation: Ensnaring two photons utilizing a bar splitter.
2. Bell-State Measurement: Playing out a joint estimation on the first photon and one of the caught photons.
3. State Reconstruction: Utilizing the estimation results to reproduce the first photon's state on the leftover ensnared photon.

This investigation gave a proof of idea to quantum instant transportation and showed the practicality of moving quantum states over distances.

The 2004 NIST Trial

In 2004, scientists at the Public Establishment of Guidelines and Innovation (NIST) transported data between two particles. This examination included:

1. Quantum Entanglement: Making ensnarement between two beryllium particles.
2. Quantum State Measurement: Estimating the condition of one particle and communicating the outcome to the area of the subsequent particle.
3. Reconstruction: Remaking the condition of the main particle on the subsequent particle utilizing the sent data.

The NIST explore denoted a huge progression by exhibiting instant transportation with material particles as opposed to photons.

The 2012 Chinese Quantum Satellite Analysis

In 2012, a group of Chinese researchers drove by Jian-Wei Container effectively magically transported the quantum condition of photons between the ground and a satellite in circle, covering a distance of north of 1,200 kilometers. This earth shattering examination included:

1. Space-based Entanglement: Making snared photon coordinates and sending one photon to the satellite while keeping the other on the ground.

2. Ground-to-Satellite Communication: Performing estimations on the ground photon and sending the outcomes to the satellite.

3. Quantum State Reconstruction: Remaking the condition of the ground photon on the satellite photon.

This trial showed the potential for significant distance quantum correspondence and laid the foundation for future quantum organizations.

Billy Carson's Outstanding Contextual investigations

Billy Carson has directed and dissected different contextual investigations to investigate the limits and utilizations of instant transportation. His work frequently consolidates hypothetical exploration with viable trial and error.

Antiquated Egyptian Instant transportation Practices

Carson's examination concerning old Egyptian texts and antiques recommends that instant transportation like peculiarities might have been perceived and polished by old human advancements. He investigates symbolic representations and compositions that portray divinities and ministers voyaging immediately between areas, recommending that these records could reflect progressed information on quantum standards.

Cognizance Driven Trials

Carson plays investigated the part of human cognizance in instant transportation through different examinations. These investigations include members endeavoring to impact quantum states utilizing centered goal and reflection. While dubious and not generally acknowledged in established researchers, Carson's work plans to overcome any barrier among cognizance and quantum mechanics.

Similar Examination with Other Instant transportation Exploration

To give a far reaching comprehension of instant transportation, contrasting Carson's discoveries and other driving exploration in the field is fundamental. This investigation features similitudes, contrasts, and possible regions for joint effort.

Quantum Processing and Instant transportation Contrasting Carson's work and headways in quantum processing uncovers likely collaborations. Quantum PCs depend on entrapment and superposition, like instant transportation. Incorporating experiences from the two fields could speed up the improvement of useful instant transportation innovations.

Otherworldly and Philosophical Methodologies

Carson's accentuation on the otherworldly parts of instant transportation diverges from the transcendently experimental focal point of standard examination. This interdisciplinary methodology energizes a more extensive point of view, taking into account the ramifications of instant transportation past the simply actual domain.

Future Headings for Instant transportation Exploration

Expanding on the triumphs and examples of past trials, future exploration can investigate new boondocks in instant transportation:

1. Scaling Up: Stretching out quantum instant transportation from particles to perceptible items, possibly in any event, living creatures.
2. Quantum Networks: Creating worldwide quantum correspondence networks utilizing satellite-based and ground-based frameworks.
3. Consciousness Integration: Examining the job of human awareness in impacting and settling quantum states.

CHAPTER 6:

Future of Teleportation

Headways in Quantum Registering and Their Effect on Instant transportation

Quantum registering, with its capacity to process and store data in quantum bits (qubits), holds the way to opening the eventual fate of instant transportation. In contrast to traditional pieces, qubits can exist in different states at the same time, because of superposition and trap. This section investigates how progressions in quantum figuring could speed up the turn of events and execution of instant transportation advances.

Quantum Incomparability and Instant transportation

Quantum matchless quality, the place where quantum PCs beat traditional PCs in unambiguous assignments, has previously been accomplished in controlled tests. As quantum figuring innovation progresses, its capacities will progressively uphold the complicated computations and continuous information handling expected for instant transportation.

Mistake Revision and Strength

One of the significant difficulties in quantum instant transportation is keeping up with soundness and limiting mistakes. High level quantum mistake rectification strategies are being created to resolve these issues. By working on the security and unwavering quality of qubits, these progressions will make huge scope instant transportation more attainable.

Combination with Quantum Organizations

The combination of quantum instant transportation with arising quantum networks is a urgent step towards viable applications. Quantum organizations, using snared particles for secure correspondence, can give the framework expected to help instant transportation on a worldwide scale. Future quantum web conventions will probably integrate instant transportation as a center part, empowering prompt information move and correspondence.

Potential for Plainly visible Instant transportation

While current investigations fundamentally include particles and photons, a definitive objective is to accomplish perceptible instant transportation, including objects and perhaps in any event, living creatures. This segment inspects the hypothetical and specialized difficulties, as well as the potential forward leaps that could make naturally visible instant transportation a reality.

Beating Scale Constraints

Increasing from individual particles to naturally visible articles includes critical difficulties, for example, keeping up with rationality across huge frameworks and dealing with the immense measures of data expected to depict complex quantum states. Propels in quantum figuring, materials science, and entrapment methods will be basic in defeating these restrictions.

Organic Contemplations

Magically transporting living creatures presents extra intricacies, including guaranteeing the uprightness and usefulness of organic frameworks. Examination into quantum science, which investigates the job of quantum processes in living organic entities, could give experiences into how to securely and successfully magically transport natural matter.

Mix with Man-made brainpower

Man-made brainpower (artificial intelligence) can assume an essential part in improving and dealing with the intricacies of instant transportation. By incorporating man-made intelligence with quantum advances, we can foster refined calculations to control and improve instant transportation processes.

Prescient Models
Artificial intelligence can be utilized to foster prescient models that expect and address blunders progressively, working on the precision

and dependability of instant transportation. AI calculations can break down immense measures of quantum information to distinguish designs and upgrade instant transportation conventions.

Robotization and Control
Simulated intelligence driven mechanization can smooth out the instant transportation process, from instating snare to overseeing information move and reproduction. Independent frameworks outfitted with man-made intelligence can deal with the complicated undertakings expected for instant transportation, diminishing the potential for human blunder and expanding proficiency.

Philosophical and Moral Ramifications
The improvement of instant transportation innovations brings up significant philosophical and moral issues. This segment investigates the more extensive ramifications of instant transportation, taking into account how it could reshape how we might interpret personality, reality, and society.

Character and Progression

Instant transportation challenges our customary thoughts of personality and progression. In the event that an individual is magically transported, is the reproduced person at the objective genuinely equivalent to the first? Logicians and researchers should wrestle with inquiries regarding cognizance, selfhood, and the idea of presence.

Moral Worries

The potential for abuse of instant transportation innovation requires hearty moral rules and guidelines. Issues like security, assent, and impartial access should be addressed to guarantee that instant transportation benefits humankind overall. Furthermore, the ramifications of magically transporting living creatures, including potential wellbeing chances and moral quandaries, require cautious thought.

Cultural Changes

The far and wide reception of instant transportation could prompt groundbreaking changes in the public eye. This segment estimates on how instant transportation could reshape different parts of human existence, from metropolitan preparation and worldwide travel to social trade and monetary frameworks.

Metropolitan and Country Elements

Instant transportation could diminish the need for urbanization by permitting individuals to live farther from their working environments, possibly rejuvenating country regions and easing metropolitan clog. It could likewise change the elements of city arranging, with instant transportation center points becoming focal highlights.

Globalization and Social Trade

Quick travel would work with more prominent social trade and grasping, separating obstructions and cultivating a more interconnected worldwide local area. It could likewise improve global participation in fields like science, schooling, and artistic expression.

Monetary Effects

Instant transportation can possibly upset customary monetary models by diminishing transportation costs, streamlining supply chains, and making new ventures. Nonetheless, it could likewise prompt work removal and require critical changes in different areas.

The fate of instant transportation holds tremendous commitment, with headways in quantum processing, artificial intelligence mix, and potential perceptible applications making ready for progressive changes. Nonetheless, it additionally presents huge difficulties and moral contemplations that should be painstakingly explored. As we draw nearer to understanding the fantasy of instant transportation, Billy Carson's bits of knowledge and interdisciplinary methodology will proceed to motivate and direct us, guaranteeing that this groundbreaking innovation is grown capably and to benefit humankind. The accompanying sections will dig into the philosophical and profound components of instant transportation, investigating its more extensive effect on how we might interpret reality and presence.

CHAPTER 7:

Philosophical and Spiritual Dimensions

The Idea of The real world and Presence

Instant transportation, particularly when seen from the perspective of quantum mechanics, challenges our conventional comprehension of the real world and presence. By separating and reassembling items or in any event, living creatures at a quantum level, transportation drives us to reevaluate existing in a specific reality.

Quantum Reality

Quantum mechanics uncovers a world that is probabilistic as opposed to deterministic. In this unique circumstance, transportation includes the reconfiguration of quantum states, recommending that items and creatures are characterized more by their enlightening substance than by their actual structure. This brings up issues about the idea of the truth: Is reality a consistent, objective element, or is it developed second by second in view of quantum data?

Character and Selfhood

Instant transportation's capability to ship an individual starting with one area then onto the next brings up significant issues about personality and selfhood. On the off chance that an individual is dismantled and reassembled somewhere else, is the magically transported individual equivalent to the first? This situation challenges our ideas of individual personality, coherence of awareness, and the pith of oneself. Logicians should wrestle with whether personality is attached to a persistent actual presence or to the coherence of data and cognizance.

Profound Ramifications

Numerous profound practices recommend that the truth is interconnected and that cognizance assumes a basic part in molding the universe. Instant transportation, by using quantum trap and the transaction of data, may offer a logical reason for these profound ideas.

Interconnectedness

Instant transportation embodies the interconnected idea of the universe, where particles can impact each other quickly over immense distances. This lines up with profound convictions that all creatures and peculiarities are interconnected. Such a viewpoint energizes a comprehensive perspective on presence, where each activity and thought has sweeping ramifications.

Awareness and Instant transportation
A few specialists, including Billy Carson, investigate the possibility that human cognizance could impact or work with instant transportation. This view sets that cognizance itself may be a basic part of the real world, equipped for communicating with quantum states. If valid, it recommends that instant transportation isn't simply an actual cycle yet additionally a cognizant one, where the psyche assumes a vital part in forming the result.

Old Insight and Current Science
Billy Carson's work frequently coordinates old insight with current logical standards. Numerous old societies, like the Egyptians and Tibetans, have fantasies and lessons about instant transportation and comparative peculiarities. Investigating these old texts can give significant

experiences and motivation to contemporary examination.

Old Texts and Practices

Old texts from different societies depict instant transportation like peculiarities, where gods or edified creatures travel immediately between areas. These records might mirror a profound comprehension of the standards basic instant transportation, yet communicated in emblematic or legendary terms. By reconsidering these antiquated stories from the perspective of quantum mechanics, we can reveal likely procedures and information that have been lost or ignored.

Incorporating Insight Customs

Present day science and old insight are many times seen as restricting methodologies, however coordinating them can prompt a more thorough comprehension of instant transportation. Antiquated practices like reflection, representation, and energy work could offer useful strategies to impact and settle quantum states, supplementing mechanical headways.

Moral and Moral Contemplations

The turn of events and execution of instant transportation advancements bring up huge moral and moral issues. Addressing these worries is critical to guarantee that instant transportation is utilized dependably and to assist all.

Protection and Assent

Instant transportation might actually disregard security and independence on the off chance that people are shipped without their assent or on the other hand assuming instant transportation gadgets are utilized for reconnaissance. Moral rules should be laid out to safeguard people's privileges and guarantee that instant transportation is utilized morally.

Impartial Access

Guaranteeing that instant transportation innovation is open to all, paying little mind to financial status, is a basic moral issue. Without cautious guideline, transportation could compound existing disparities, giving favored bunches an unreasonable benefit.

Influence on Society

Instant transportation could disturb cultural designs and standards, prompting unseen side-effects. For instance, the capacity to travel promptly could influence populace appropriation, work markets, and social elements. Expecting and dealing with these effects is fundamental to forestall adverse results.

Philosophical Psychological studies

Investigating philosophical psychological studies can assist us with figuring out the ramifications of instant transportation and challenge our suppositions about the real world, personality, and profound quality.

The Instant transportation Mystery

Envision an instant transportation gadget that makes a precise duplicate of an individual at the objective while obliterating the first. Is the individual who shows up at the objective equivalent to the person who withdrew? This psychological test questions the idea of character and progression, moving us to rethink being a similar individual.

The Boat of Theseus

The Boat of Theseus is an exemplary philosophical psychological study that investigates the idea of character. On the off chance that a boat's parts are steadily supplanted until none of the first parts remain, is it as yet unchanged boat? Applied to instant transportation, this examination brings up issues about whether an individual who is magically transported, with each molecule reassembled, is a similar individual as in the past.

CHAPTER 8:

The Intersection of Science Fiction and Reality

Sci-fi's Effect on Instant transportation Exploration

Sci-fi has long roused researchers and pioneers, molding public discernment and driving interest in instant transportation. Notorious works from writing, film, and TV have given creative dreams of instant transportation, frequently foreseeing mechanical headways and investigating their cultural ramifications.

Writing

From H.G. Wells' "The Time Machine" to Isaac Asimov's "Establishment" series, sci-fi writing has habitually investigated subjects of instant transportation. These accounts engage as well as incite pondered the potential and traps of moment travel.

- "The Hike" by Stephen King: In this brief tale, Ruler investigates the mental and actual results of instant transportation, featuring likely risks and moral situations.
- "The Stars My Objective" by Alfred Bester: This novel presents "jaunting," a type of individual instant transportation, and inspects how such an innovation could modify society and individual way of behaving.

Film and TV
Well known movies and television series have carried the idea of instant transportation to a wide crowd, frequently portraying it as a standard innovation in cutting edge social orders.

- "Star Trek": The notorious carrier innovation utilized by the Starship Endeavor team has become inseparable from instant transportation. It has roused numerous researchers to investigate the potential outcomes of issue move.
- "The Fly": This film offers a useful example about the unanticipated outcomes of instant

transportation, zeroing in on the potential for devastating mistakes simultaneously.

Genuine Motivations from Fiction

Numerous researchers and designers acknowledge sci-fi as a wellspring of motivation for their work on instant transportation. The creative situations portrayed in fiction urge analysts to think past the ongoing impediments and investigate additional opportunities.

Quantum Trap and Sci-fi

The idea of quantum trap, a critical part of instant transportation, has been promoted by sci-fi. This has assisted with raising public mindfulness and interest, prompting expanded subsidizing and research in quantum mechanics.

- Public Discernment and Funding: The depiction of instant transportation in sci-fi has helped accumulate public help for research in quantum mechanics and related fields, impacting strategy and subsidizing choices.

Overcoming any barrier: From Fiction to Plausibility

Interpreting the creative ideas of sci-fi into reasonable innovation includes beating critical logical and specialized difficulties. This segment investigates the present status of exploration and the means expected to make instant transportation a reality.

Mechanical Obstacles

While sci-fi frequently overlooks the specialized subtleties, genuine instant transportation requires taking care of intricate issues connected with quantum data, trap, and mistake remedy.

- Information Transfer: Transportation requires the exchange of huge measures of quantum data, which should be finished with outrageous accuracy to guarantee exactness.

- Dependability and Coherence: Keeping up with the intelligibility of quantum states over significant distances and times is a significant test, requiring progressed materials and procedures.

Interdisciplinary Exploration
Making instant transportation a reality will require joint effort across different logical disciplines, including physical science, software engineering, designing, and science.

- Quantum Registering and Communication: Advances in quantum figuring and secure correspondence networks are fundamental for creating reasonable instant transportation advances.
- Natural and Clinical Research: Understanding the impacts of instant transportation on organic entities is pivotal, especially assuming we plan to magically transport living creatures.

Expected Leap forwards and Advancements

A few arising innovations and hypothetical progressions could carry us closer to accomplishing instant transportation.

Quantum Instant transportation Trials

Ongoing examinations have exhibited the attainability of quantum instant transportation over expanding distances, recommending that viable applications may not be far away.

- Significant distance Quantum Networks: Creating vigorous quantum networks with repeaters and high level mistake amendment could empower worldwide quantum instant transportation.
- Photonic and Nuclear Teleportation: Proceeded with investigation into magically transporting photons and molecules will give the establishment to additional perplexing frameworks.

The Job of artificial intelligence and AI

Man-made brainpower and AI can assume a urgent part in dealing with the intricacies of instant transportation, from mistake remedy to framework improvement.

- Prescient Algorithms: computer based intelligence can foster prescient calculations to expect and address mistakes continuously, working on the unwavering quality of instant transportation.
- Mechanization and Control: man-made intelligence driven robotization can smooth out the instant transportation process, diminishing human mistake and expanding proficiency.

Cultural Effects and Moral Contemplations
As instant transportation moves from sci-fi to the truth, its cultural effects and moral ramifications should be painstakingly thought of.

Monetary Disturbance
Instant transportation could reform ventures like transportation, coordinated operations, and

medical services, yet it could likewise prompt work dislodging and monetary disturbance.

- New Business Models: Organizations should adjust to new plans of action based on instant transportation, possibly prompting huge monetary movements.
- Labor force Retraining: Setting up the labor force for the progressions brought by instant transportation will be fundamental, with an emphasis on retraining and training.

Protection and Security

Instant transportation innovation could present huge protection and security chances, requiring hearty administrative systems and moral rules.

- Unapproved Teleportation: Forestalling unapproved instant transportation and guaranteeing assent will be critical to safeguarding individual freedoms.

- Reconnaissance and Control: Transportation could be abused for observation or control, raising worries about security and independence.

CHAPTER 9:

Challenges and Controversies

Specialized Difficulties in Instant transportation While instant transportation vows to upset how we might interpret travel and correspondence, critical specialized obstacles should be defeated to make it a useful reality.

Quantum Intelligibility and Decoherence

One of the essential difficulties in instant transportation is keeping up with quantum cognizance. Quantum states are incredibly sensitive and can undoubtedly be disturbed by ecological variables, prompting decoherence.

- Keeping up with Coherence: Methods to protect quantum frameworks from outside obstruction are fundamental. This incorporates progressed cryogenics to keep up with low temperatures and confinement strategies to safeguard against electromagnetic obstruction.
- Blunder Correction: Creating hearty quantum mistake revision codes is basic. These codes distinguish and address blunders in quantum data without obliterating the quantum state.

Adaptability and Effectiveness
Magically transporting single particles, for example, photons, has been accomplished, however scaling this cycle to bigger frameworks and articles presents colossal difficulties.

- Entrapment Generation: Making and keeping entrapped states over lengthy separations and enormous frameworks require progressions in quantum organizing and the advancement of dependable quantum repeaters.
- Information Handling and Storage: Transportation includes handling and putting

away immense measures of quantum data. Proficient calculations and versatile quantum processors are important to deal with this information.

Moral and Moral Discussions

Instant transportation innovation presents moral and moral quandaries that should be addressed to guarantee its capable turn of events and use.

Personality and Progression

Instant transportation brings up key issues about private personality and congruity. In the event that an individual is dismantled and reassembled somewhere else, is the subsequent individual equivalent to the first?

- Philosophical Implications: Thinkers should consider whether the magically transported individual holds a similar cognizance and way of life as the first, or on the other hand in the event that a new, vague duplicate is made.
- Legitimate and Social Considerations: Overall sets of laws should address the situation with

magically transported people. Issues like privileges, obligations, and progression of personhood will require new legitimate systems.

Protection and Assent

Instant transportation innovation can possibly disregard security and independence, requiring strong moral rules and administrative measures.

- Unapproved Teleportation: Forestalling unapproved utilization of instant transportation gadgets is vital to safeguard people's security and assent.
- Observation and Control: The potential for abuse of instant transportation for reconnaissance or control raises huge moral worries. Guidelines should guarantee that instant transportation is utilized dependably and morally.

Cultural and Financial Effects

The boundless reception of instant transportation could prompt significant cultural and financial changes, both positive and negative.

Monetary Interruption

Instant transportation could reform different enterprises, including transportation, operations, and medical services, however it could likewise prompt work dislodging and monetary disturbance.

- Industry Transformation: Ventures reliant upon conventional transportation strategies might confront huge disturbance, expecting variation to new plans of action based on instant transportation.
- Labor force Retraining: Setting up the labor force for the progressions brought by instant transportation will be fundamental. This incorporates retraining projects and schooling drives to assist laborers with changing to new jobs.

Social Elements

Instant transportation could change social elements, including populace conveyance, urbanization, and relational connections.

- Metropolitan and Country Dynamics: Transportation could diminish the need for urbanization, permitting individuals to live farther from their working environments. This could renew provincial regions and ease metropolitan clog.
- Relational Relationships: Quick travel could upgrade individual connections by making it more straightforward to visit loved ones, however it could likewise challenge the thought of actual presence in keeping up with social bonds.

Logical and Mechanical Discussions
The advancement of instant transportation innovation is joined by logical and mechanical discussions, reflecting varying points of view and needs.

Quantum Translations
Various translations of quantum mechanics, like the Copenhagen understanding, many-universes translation, and pilot-wave hypothesis, offer

differing viewpoints on the possibility and nature of instant transportation.

Copenhagen Interpretation: This understanding proposes that quantum states breakdown upon estimation, bringing up issues about how instant transportation can protect cognizance.
- Many-Universes Interpretation: As per this translation, transportation could include expanding into equal universes, which has significant ramifications for how we might interpret reality.

Specialized Possibility

The specialized possibility of instant transportation stays a subject of discussion among researchers and designers, with contrasting perspectives on the timetables and techniques for accomplishing down to earth instant transportation.

- Hopeful Projections: A few scientists accept that progressions in quantum registering and systems administration will empower down to

earth instant transportation inside years and years.
- Doubtful Views: Others contend that the specialized difficulties are too vital for even think about conquering soon, requiring leap forwards in central physical science and designing.

Public Insight and Acknowledgment
Public discernment and acknowledgment of instant transportation innovation will assume a urgent part in its turn of events and execution.

Social Impacts
Social perspectives towards instant transportation, molded by sci-fi and well known media, impact public insight and acknowledgment.

- Positive Reception: Positive depictions of instant transportation in media can create fervor and backing for innovative work.
- Dread and Skepticism: Negative depictions, featuring expected risks and moral difficulties,

can prompt apprehension and wariness, influencing public acknowledgment.

Training and Effort

Compelling training and effort drives are fundamental to illuminate the general population about the advantages, chances, and moral contemplations of instant transportation innovation.

CONCLUSION

Instant transportation remains at the crossing point of significant logical potential and philosophical request, overcoming any issues between the domains of sci-fi and reality. As we've investigated all through this book, Billy Carson's experiences into instant transportation enlighten both the tempting prospects and the imposing difficulties of this extraordinary innovation.

Summing up Key Experiences

Establishments and Quantum Insights:
We started by analyzing the essential standards of instant transportation, grounded in quantum mechanics and entrapment. Billy Carson's work highlights the meaning of these standards, stressing the requirement for proceeded with research in quantum cognizance and mistake adjustment to conquer specialized hindrances.

Genuine Applications and Future Prospects:
We investigated the down to earth uses of instant transportation, from immediate correspondence and secure information move to progressive changes in transportation and strategies. The fate of instant transportation depends on progressions in quantum registering, computer based intelligence combination, and conquering versatility issues.

Philosophical and Otherworldly Dimensions:

Instant transportation's suggestions stretch out past innovation, testing how we might interpret reality, personality, and presence. The philosophical and profound aspects brief us to reexamine our ideas of selfhood, interconnectedness, and the job of cognizance in forming reality.

Sci-fi and Reality:
Sci-fi has long motivated logical investigation, offering creative dreams of instant transportation that drive advancement. Overcoming any barrier among fiction and reality includes tending to critical specialized difficulties, informed by the imaginative potential outcomes portrayed in writing, film, and TV.

Challenges and Controversies:
Instant transportation's improvement is laden with moral and moral quandaries, from inquiries of personality and coherence to worries about protection, assent, and cultural effects. Tending to these difficulties requires interdisciplinary

joint effort, hearty moral rules, and insightful thought of public discernment.

The Way ahead

As we plan ahead, the way ahead for instant transportation includes a cautious equilibrium of logical development, moral obligation, and cultural preparation. Key areas of center include:

Interdisciplinary Research:
Proceeded with joint effort across physical science, software engineering, designing, and science is crucial for advance comprehension we might interpret quantum states and foster commonsense instant transportation innovations.

Moral Frameworks:
Laying out thorough moral structures and administrative rules will guarantee that instant transportation is utilized dependably, safeguarding individual freedoms and tending to cultural effects.

Public Engagement:
Viable instruction and effort drives are pivotal to cultivate informed popular feelings and backing. Drawing in different partners in the discussion about instant transportation will assist with forming a future where this innovation helps all of mankind.

Philosophical Inquiry:
Continuous philosophical investigation into the idea of the real world, personality, and awareness will advance comprehension we might interpret instant transportation's more extensive ramifications, directing moral and capable turn of events.

Last Reflections

Billy Carson's interdisciplinary way to deal with instant transportation, incorporating old insight with state of the art science, offers a one of a kind and comprehensive point of view on this extraordinary innovation. As we keep on investigating the conceivable outcomes and difficulties of instant transportation, his bits of knowledge help us to remember the significance of offsetting development with moral obligation and philosophical reflection.

Instant transportation can possibly reshape our reality in significant ways, from upsetting correspondence and transportation to testing our key comprehension of presence. By exploring the logical, moral, and philosophical components of instant transportation nicely and cooperatively, we can outfit its groundbreaking power to ultimately benefit humankind.

APPENDICES

Reference section A: Key Ideas in Quantum Mechanics

Quantum Entanglement: A peculiarity where particles become interconnected with the end goal that the condition of one molecule promptly impacts the condition of another, no matter what the distance between them.

Quantum Coherence: The property of quantum states to show obstruction impacts, keeping an obvious stage relationship.

Quantum Decoherence: The cycle by which a quantum framework loses its lucidness, frequently because of collaboration with the climate, bringing about the progress from a quantum to a traditional state.

Quantum Superposition: The rule that a quantum framework can exist in different states at the same time until it is estimated, so, all in all it implodes into one of the potential states.

Quantum Teleportation: The most common way of sending the specific condition of a quantum framework starting with one area then onto the next without actual exchange, depending on snare and old style correspondence.

Addendum B: Verifiable Achievements in Instant transportation Exploration

1993: Charles Bennett and his group at IBM propose the hypothetical structure for quantum instant transportation, showing that magically transporting quantum information is hypothetically conceivable.

1997: Trial showing of quantum instant transportation by a group drove by Anton Zeilinger, transporting the condition of a photon over a brief distance.

2003: Further headways in quantum instant transportation, including magically transporting the condition of particles and particles, growing the extent of instant transportation research.

2012: Fruitful quantum instant transportation over a distance of 143 kilometers between the

Canary Islands, denoting a huge accomplishment in significant distance quantum correspondence.

2020: Specialists accomplish quantum instant transportation between two central processors interestingly, featuring progress in coordinating instant transportation with quantum registering.

Index C: Moral Rules for Instant transportation Innovation

1. Informed Consent: Guaranteeing that people completely get it and agree to the course of instant transportation, including expected dangers and suggestions.

2. Privacy Protection: Carrying out measures to safeguard people's security and forestall unapproved utilization of instant transportation innovation.

3. Equitable Access: Guaranteeing that instant transportation innovation is open to all, paying

little heed to financial status, to forestall fueling existing imbalances.

4. Safety Protocols: Laying out thorough security conventions to limit the gamble of damage during the instant transportation process.

5. Legal Frameworks: Creating legitimate structures to resolve issues of personality, coherence, and obligation with regards to instant transportation.

Index D: As often as possible Got clarification on some things (FAQs)

Q1: Is instant transportation at present possible?
A1: Quantum instant transportation of data has been tentatively illustrated, however magically transporting bigger items or living creatures stays a critical test and isn't yet doable.

Q2: How does quantum instant transportation work?

A2: Quantum instant transportation includes trapping two particles, sending the quantum condition of one molecule to the next through traditional correspondence, and utilizing the entrapment to reproduce the first state.

Q3: What are the fundamental difficulties in creating instant transportation technology?
A3: Key difficulties incorporate keeping up with quantum lucidness, scaling the interaction to bigger frameworks, creating productive information handling calculations, and tending to moral and cultural ramifications.

Q4: What are the possible uses of teleportation?
A4: Potential applications incorporate secure quantum correspondence, moment information move, high level processing organizations, and groundbreaking changes in transportation and operations.

Q5: What are the moral worries related with teleportation?

A5: Moral worries incorporate issues of personality and congruity, security and assent, impartial access, and potential cultural effects like financial disturbance and changes in friendly elements.

Informative supplement E: Proposed Further Perusing

Books:
- "Quantum Calculation and Quantum Data" by Michael A. Nielsen and Isaac L. Chuang
- "The Texture of The real world" by David Deutsch
- "The Secret Reality: Equal Universes and the Profound Laws of the Universe" by Brian Greene

Articles:
- "Exploratory Quantum Instant transportation" by Dik Bouwmeester et al., Nature

- "Quantum Ensnarement and Instant transportation" by Anton Zeilinger, Logical American
- "Instant transportation Physical science Study" by Eric W. Davis, Flying corps Exploration Research facility

Websites:
- [Quantum Data and Computation](https://quantum-computation.ibm.com)
- [Public Organization of Guidelines and Innovation (NIST) Quantum Data Program](https://www.nist.gov/programs-projects/quantum-data)

Finish of Supplements

These addendums give fundamental data and assets to additional investigation of instant transportation, its logical establishments, moral contemplations, and momentum research. By digging into these materials, perusers can develop how they might interpret the intricacies and capability of instant transportation innovation, directed by the bits of knowledge and interdisciplinary methodology advocated by Billy Carson.

ACKNOWLEDGEMENTS

Composing this book has been an excursion of investigation and revelation, and it could never have been conceivable without the help, experiences, and commitments of numerous people and associations.

Above all else, I might want to offer my most profound thanks to Billy Carson. His visionary thoughts and significant experiences into instant transportation have been the foundation of this work. His obligation to investigating the crossing points of science, otherworldliness, and old insight has propelled me to look past traditional limits and embrace an all encompassing way to deal with figuring out instant transportation.

I'm colossally appreciative to the specialists and researchers who have devoted their lives to propelling comprehension we might interpret quantum mechanics and instant transportation. Their noteworthy examinations and hypothetical work have made ready for future developments and gave a strong groundwork to this book.

Unique much gratitude goes to my loved ones for their immovable help and consolation all through the creative cycle. Their understanding, and confidence in my vision have been priceless.

I might likewise want to recognize the commitments of the accompanying people and associations:

- Dr. Anton Zeilinger and his examination team: For their spearheading work in quantum instant transportation and their ability to share their insight and discoveries.

- The Quantum Figuring and Data community: For encouraging a climate of cooperation and development, which has been critical for the advancement in this field.
- Editors and Companion Reviewers: For their careful audit and helpful criticism, which have fundamentally worked on the quality and lucidity of this book.

At long last, I might want to thank my perusers. Your interest and interest in the captivating prospects of instant transportation have persuaded me to dig further and investigate this subject with enthusiasm and devotion. I trust this book rouses you to envision, question, and add to the continuous discourse about the fate of instant transportation.